RAINMAKERS
Existence's Blueprint

A deep understanding of Earth's design is fundamental to solving our greatest challenges.

From
Leeku's Environmental
Observation Book Series

Mastewal E. Ademe

Copyright © 2024 Mastewal E. Ademe

ISBN: 9798340885951

All rights reserved. No part of this publication may be reproduced, distributed, or transmitted in any form or by any means, including photocopying, recording, or other electronic or mechanical methods, without the prior written permission of the publisher, except in the case of brief quotations embodied in critical reviews and certain other noncommercial uses permitted by copyright law.

Printed in the United States of America.

Water evaporates and then moves through a continuous cycle, which involves extensive phases of transfer, redistribution, and precipitation.

This extensive and complex process is often oversimplified as "the water cycle," overlooks the complexity, scale and significance of this process to life.

The world is intricately designed to manage this immense workload: humans alone cannot do the job needed.

Read this book to explore the complete system and understand how it all functions to sustain life and to figure out our roles thereby feeding the future and improving climate.

The future holds immense promise if we embrace and implement the principles outlined in this book.

Earth's natural design is both magnificent and complex — the greatest mobilizer of water, yet also the greatest source of water, soil, and nutrient loss. However, with the right approach, this very design holds the greatest potential to address critical global challenges such as rising food prices, forced migration, conflict, and water scarcity.

Table of Content   Page

I Background study

1. Summary……………………………………………...…12
2. Existence's Blueprint………………………………….....18
3. The Need for Universal Unity……………………………20
4. The Water Cycle: Lifeblood of Our Planet……………..23
   a. Rain's Role in Civilization and Agriculture
   b. The Cost and Impact of Artificial Rainmaking
5. Challenges of Rain: From Flooding to Wildfires………..27
   a. Wasted Rainwater: Turning it into Opportunity
   b. Global Warming on Rainfall Patterns
   c. Mitigating Rainfall Challenges
6. Strategies for Rainwater and Climate………………....32
7. Cooling the Atmosphere………………………………36
   a. Cooler Environments and Cloud Formation
   b. Water Settlement
   c. Tree Planting for Water and Climate
8. Literature review……………………………………….41

II The Design Part

9. The Thirteen Existence's Blueprint Factors……………48
10. Design Gaps……………………………………………76

III What is required for the design to operate as envisioned?

11. The Factors That Sustain Rain…………………………80
12. Human Interferences out of the 13 Factors………...…82
13. Manaus Qualifies to get Rain but not Dubai…………..85
14. Egypt Depends on Rain………………………………...87
15. Climate Change Warning……………………...………89
16. We are Lucky to be in this World…………………...…91
17. The Way Forward……………………………………...92

18. Engaging and sharing..97
19. Recommendation..99
20. References..102

The Author's Revelation About Rain

Through the journey of writing the *Leeku Environmental Observation Book Series* for children, the author experienced a profound awakening: a deeper understanding of rain's central role in sustaining life on Earth.

By simplifying complex environmental and climate science for young readers, the author uncovered not only the elegance of nature's systems but also the brilliance of rainfall as a global water transport mechanism. Rain emerged as more than just weather — it became clear as a finely tuned, natural engineering system responsible for gathering, purifying, distributing, and renewing the planet's water supply.

Each rainfall is a product of a delicate interplay between multiple environmental factors. When any of these are disrupted due to land degradation, deforestation, or emissions the balance is lost, and the water cycle falters. From this realization, the author identified 13 critical factors required for rain to occur, emphasizing the importance of protecting and promoting them to address pressing global challenges such as food security, water scarcity, and climate instability.

This journey revealed an undeniable truth: rain is not accidental, it is part of Earth's intentional, life-sustaining design. By understanding and preserving the systems that make rain possible, society can play an active role in restoring climate balance and ensuring resilience for generations to come.

The Author's Message

After years of exploring rainwater management from the ground up, my understanding deepened beyond infrastructure or policy. I came to a life-defining realization: life is governed by a single law: the principle of continuous supply.

From a single cell to the entire planet, life depends on the uninterrupted availability of essentials — oxygen, food, and water. While we all recognize the critical need for oxygen, the equal importance of continuous water and food supply is often overlooked.

Our bodies are designed for this balance. Water occupies less space than oxygen, so it is stored throughout our tissues and bloodstream. Yet, even with built-in reserves, dehydration begins to threaten life within just a few days. This is not random — it is intentional design.

This same principle scales the Earth itself. Rainfall, and the mechanisms that support it, follow this blueprint ensuring that water reaches plant roots, sustains ecosystems, and supports biodiversity. Earth's systems are not merely functional; they are designed to preserve life through constant provision.

Before we tackle climate change, we must first understand this underlying structure. When we recognize that Earth is purposefully designed to support life through continuous cycles — particularly of water — we move from fragmented fixes to informed, unified action.

Rainmakers: Existence's Blueprint is an invitation: to leaders, changemakers, and organizations. Once we understand nature's design, we can become active participants in healing it.

I

Background study

1. Summary

One of the fundamental principles of life is continuous supply. Without it, life cannot be sustained. In essence, life itself is defined by the uninterrupted availability of essential resources: among which water stands as one of the most vital.

More emphasis is very important, just as we rely on a continuous supply of oxygen to survive, we are equally dependent on a constant supply of water. Vegetation, which forms the basis of our primary food sources, requires an uninterrupted flow of water through its roots to thrive. Without it, plants wither and die, breaking the chain of life. This underscores a critical truth: our survival is inseparably linked to the continuous availability of water, just as it is to oxygen.

In other words, the mere presence of water is not enough to sustain life; it must be replenished consistently to meet the demands of ecosystems, agriculture, and all animals. Without regular rainfall, humans lack the capacity to ensure a global and equitable water supply. Rain is not only a natural miracle but also an irreplaceable mechanism for sustaining life on Earth, far beyond what human ingenuity can replicate at scale.

Even though all water resources on land surfaces depend on rain; today, human-induced irrigation provides just a small fraction of less than 1% of the water that rainfall delivers to agriculture and ecosystems worldwide.

This emphasizes that rain is by far the largest and most essential source of water for food production, upon which both humans and animals completely depend on for nourishment and hydration. Rain itself is a product of the water cycle, an intricate system that replenishes our rivers, soils, and ecosystems, supporting life across the planet. On the other hand, without rain, there would be no fresh water. Ocean water is undrinkable, and all other surface sources eventually stagnate or dry up.

Existence's Blueprint as a metaphor for the intricate natural systems that govern our world, emphasizing the need to understand and respect these principles for long-term sustainability. The book highlights how the world is designed to provide us the rainwater with areas where we can avoid damage and/or make improvements, and it offers practical solutions with principles to follow.

It advocates for a unified approach to water management that empowers communities, to secure their water resources and build climate resilience.

Readers will gain valuable insight into critical factors that provide us with rainwater every year, making this book an essential guide for anyone looking to understand and support the natural systems that sustain life. These topics encourage communities to exchange ideas, spark dialogue, and learn from one another.

Because it emphasizes the crucial role of the rain process within the larger environmental system, highlighting the need for a deeper understanding of its intricacies. Although rain is essential for sustaining life, it often receives inadequate attention, even though human activities frequently disrupt the delicate balance of the water cycle.

It explores the concept of "No Rain, No Life" and underscores how global warming impacts the water cycle, which regulates weather patterns, ecosystems, and global temperatures. To achieve meaningful climate improvements, it is essential to recognize both the challenges and the opportunities for positive intervention, ensuring that human actions are managed in an environmentally sustainable way.

Rain is vital to life, driving essential processes like food production, clean water supply, and climate regulation. On top of that rain cleans the air and land surfaces from dirt leaving smaller areas flooded with dirt including rivers, ponds and lakes.

The water cycle operates on a massive scale, moving an enormous volume of water globally, a process impossible for humans to replicate. However, the remarkable complexity of water mobilization remains largely undervalued and overlooked.

However, the water cycle is no longer functioning efficiently across the globe. Increasingly erratic rainfall occurrences and more frequent heavy floods suggest that the system is not operating according to its original design.

Effective rain management begins with education. Teaching water cycle is essential, not only for addressing climate change but also for ensuring a sustainable future for the next generation. By understanding how Earth's design functions and how global warming impacts water resources, communities can adopt smarter, more sustainable practices that protect the environment and secure long-term resilience.

Raising awareness can drive climate action and promote environmental stewardship. Knowledge about the water cycles design can lead to effective strategies that address issues like droughts, flooding, and water scarcity, helping to create a more resilient ecosystem.

The objective of this book is to inform communities about the design of rain, the principles that we must follow and empower each of us to contribute to climate improvement and increase food production for a sustainable future.

Rainfall is essential not only for domestic water supply but also for agriculture and construction; those make it as the foundation of human survival. Despite its importance, the broader implications of the water cycle are often overlooked in education and public discourse.

As stated above just as oxygen is vital without interruption, a continuous supply of water is crucial for both animals and plants. Rainfall is fundamental to maintaining the hydrological balance in ecosystems with the intention that ensures continuous water supply. To achieve its goal, it plays a crucial role in recharging groundwater aquifers, filling rivers and lakes, and maintaining soil moisture with predetermined timing i.e. as to the design.

This stored fresh water acts as a natural buffer, supporting plant growth, sustaining wildlife, and ensuring the availability of water during dry periods. These natural **storage facilities** regulate the water supply, which is crucial for the resilience of ecosystems and human communities alike. By protecting, upgrading and managing these storage resources wisely, we can better cope with seasonal variations and climate change impacts.

While rain is essential, it can also bring challenges. Heavy rainfall can lead to flooding, property damage, and soil erosion. Conversely, a lack of rain can cause droughts, leading to water shortages, agricultural decline, and wildfires. Maintaining balanced rainfall is critical to preventing environmental and economic damage.

Out of the necessities mentioned earlier, air is easily accessible, but unlike air, fresh water is limited and must be carefully managed to ensure continuous supply. Various scientists have emphasized the critical importance of regulating the water cycle for our survival.

Atmospheric moisture influences weather patterns and understanding how to harness this water can help manage resources and mitigate extreme weather events. By capturing airborne moisture, we can turn it into a valuable resource, enhancing our ability to handle both water scarcity and excess.

Fig. Nile River fall in Ethiopia locally known as "Tiss-Issat"

2. Existence's Blueprint

As outlined above, rain is the primary source of fresh water for all land-based life. However, the mere presence of water is not enough to sustain life—it must flow through natural cycles that purify, renew, and distribute it continuously. The purpose of the water cycle, especially rainfall, is to provide a consistent and balanced supply of fresh water in the right quantities throughout the year, supporting ecological stability. With this understanding, the importance of managing water on a global scale becomes clear.

The solar power and the Earth's rotation, from the grandest planetary scale down to the movement of the tiniest water vapor molecule, serves a deliberate purpose. Every element in this world is intricately designed to sustain and nurture life. After reading this book, you will gain a deeper understanding of the fundamental blueprint of existence: uncovering the natural principles that govern our world and how they contribute to its remarkable capacity to support life.

It suggests the foundational design underlying all life and systems in the universe. It evokes the idea of a structured, purposeful pattern or set of principles that guide the way life is sustained and interacts with its environment.

In this sense, **Existence's Blueprint** points to both the natural laws governing the planet and the human responsibility to respect and align with these laws for long-term sustainability.

No rain, no life. This compelling book explores the critical role rainwater plays in sustaining life, ecosystems, and agriculture, while revealing the often-overlooked importance of the water cycle in climate discussions.

The book emphasizes indirectly how human activities like deforestation, urbanization, manufacturing, and unsustainable farming disrupt natural processes, leading to droughts, floods, and climate instability. It presents practical, science-backed principles for solutions, including importance of rainwater knowledge, cooling the air, enhancing soil cover, minimizing runoff, avoiding soil erosion and increasing recharging.

By merging traditional knowledge with modern innovations, this book advocates for a global shift in water management and urges communities worldwide to adopt sustainable practices. Readers will learn essential factors for understanding how the universe works to provide us with rainwater. The knowledge gained help in boosting community participation with environmental efforts, emphasis on empowering people in developing countries to secure their water future.

3. The Need for Universal Unity

While detailed studies have been conducted on climate, the rain process remains viewed as a natural and delicate phenomenon, often without a full understanding of its intricacies. Though it is a natural process, it is essential to recognize that it is a system of purposeful design for our existence. It is our responsibility to comprehend how this process works so that we can find ways to enhance it.

The aim is to provide a holistic explanation of how rain functions within the larger environmental system and identifies our role in ensuring that our consumption and interventions are sustainable and environmentally friendly. It explores the critical role of the water cycle in sustaining life and supporting climate resilience.

Human activities frequently disrupt the delicate balance of the water cycle, yet within these challenges lie numerous opportunities for positive intervention. If we fail to recognize and responsibly address both the consequences of these disruptions and the potential solutions, our efforts to mitigate climate change will fall short.

A comprehensive understanding of the water cycle affected by global warming is crucial for achieving the cumulative impact needed for meaningful climate improvements. The water cycle not only regulates weather patterns but also maintains ecosystems and balances global temperatures. Effective climate action demands an understanding of how water moves through the atmosphere, soil, and oceans and how human activities alter and destabilize this essential system.

Fig. Clouds: taken at Laurel, Maryland USA

The atmosphere is a shared resource, transcending geographical boundaries. Positive environmental actions in one location can benefit other regions far beyond national borders, emphasizing the need for global cooperation in climate improvement.

Unity in addressing climate challenges is crucial, as atmospheric conditions and their impacts do not adhere to political boundaries and affect everyone globally. We all share the same air, and the same water flows through our bodies, our blood, and our cells. In life, we are deeply connected, and in death, we return to the same soil. It is time to awaken to this reality: we are not separate from nature, but part of one living system.

4. The Water Cycle: Lifeblood of Our Planet

Water is fundamental to life, but while many are familiar with the basic water cycle evaporation, condensation, and rainfall few grasp its full complexity. **No one asks how the water cycles.** The water cycle is often simplified to just the processes driven by the sun's heat. While solar energy is a key factor, the full mechanism of the water cycle is more complex, involving several interconnected processes, forces, characters, size, setup, energies and resources that influence global water distribution, ecosystems, and climate.

Various studies have extensively highlighted the critical role of the water cycle in sustaining and accommodating life on Earth. The continuous movement of water between the atmosphere, land, and oceans supports plant growth, replenishes freshwater sources, and moderate temperatures (Jones et al., 2023). It creates the conditions necessary for agriculture, sustains ecosystems, and maintains water availability. A disruption in the water cycle would lead to ecosystem collapse, water scarcity, food insecurity, and extreme weather conditions, making the planet inhospitable.

Our survival is intricately connected to this natural process. Hence understanding this topic should be one of our missions. Although water is abundant on Earth, its cyclical availability through rain is what sustains ecosystems. This emphasizes that life depends not just on the presence of water, but on its regular distribution (USGS, 2023). At this point, again it implies no rain, no life.

The water cycle functions as a vast, self-sustaining global system. The sheer volume of water moved across the planet is immense, and replicating this cycle artificially would be impossible, both in terms of scale and cost (Cao et al., 2022). Yet, this life-sustaining system is provided freely by nature, benefiting all living organisms. Mismanagement of these resources, however, has led to environmental degradation, which, if reversed, could contribute to a more sustainable world.

For example, in a specific district, estimating runoff and evaporation can reveal the amount of lost water. If we convert this loss into monetary value, it could provide the resources needed to develop the area sustainably, reducing reliance on outside assistance.

A. Rain's Role in Civilization and Agriculture

Rain not only sustains ecosystems but also shapes human civilization. Rivers, which have nurtured societies throughout history, rely on consistent rainfall. Oceans, while vast, contain undrinkable water; it is rain that purifies and replenishes freshwater sources, ensuring water for agriculture and human survival. Since rivers depend on rain implies irrigation practices depend on rainwater, making rain essential for agriculture and life itself (Mekonnen & Hoekstra, 2021).

The water cycle, driven by rainfall, ensures the necessary supply of water for drinking and food production. Rainfed agriculture is the primary source of food production in many countries globally, relying on natural rainfall rather than irrigation to sustain food supply.

This practice supports the livelihoods of millions, making it a critical component of global food security. However, its success is closely tied to climate conditions, making it vulnerable to unpredictable rainfall patterns and climate change. Hence rainwater is not just a natural resource; it is the foundation of life and civilization, and managing it wisely is crucial for our planet's future.

B. The Cost and Impact of Artificial Rainmaking

While natural rainwater comes without a monetary cost, artificial rainmaking used to supplement water supplies in times of scarcity can carry a significant financial burden. The infrastructure, ongoing maintenance, and operational costs of artificial methods are substantial, particularly in smaller areas where the return on investment may not be as efficient. These costs highlight the importance of preserving natural water cycles, which remain critical to sustainable development. Relying too heavily on artificial interventions can lead to unforeseen economic and environmental consequences.

Rain symbolizes nature's invaluable contribution to life, and the broader implications of artificial rainmaking need to be carefully considered. While it may address immediate shortages, the long-term impacts on ecosystems and regional climates require thorough study. As we explore these concepts, we recognize the critical importance of maintaining natural rainfall and focusing on sustainable water management.

This topic emphasizes the value of natural systems and the need to align human practices with environmental principles. By framing rainwater as a freely provided natural resource—and contrasting it with the high costs of artificial water systems like drilling, pumping, treatment, and distribution—it makes a compelling case for adopting sustainable rainwater management.

This comparison can serve as a powerful motivation for adopting more eco-friendly water management strategies, encouraging communities and policymakers to leverage natural processes rather than relying solely on costly and resource-intensive artificial methods.

5. Challenges of Rain: From Flooding to Wildfires

Excess rainfall

While rain is essential for life, it also presents significant challenges as mentioned earlier. Heavy rainfall can lead to flooding, causing property damage, transportation disruptions, and the displacement of communities. Prolonged rain periods may result in soil erosion, reducing land productivity and increasing agricultural challenges.

Additionally, excessive moisture fosters mold growth, threatening both human health and building structures. In mountainous or hilly areas, landslides triggered by rain pose further risks to vulnerable populations.

Erratic Rainfall

Conversely, disruptions in rainfall patterns can lead to severe droughts, and destabilizing ecosystems. Insufficient rainfall diminishes water availability, accelerates soil desiccation, and hampers crop productivity, directly impacting food security and agricultural livelihoods.

Prolonged dry conditions also foster an environment highly susceptible to wildfires, where vegetation dries out and becomes fuel for rapidly spreading fires. These wildfires can decimate forests, wildlife habitats, and human communities, often resulting in long-term ecological damage and economic loss.

Maintaining Balance

The stark contrast between periods of excessive rainfall and drought highlights the critical importance of maintaining consistent, balanced rainfall patterns, which are fundamental not only for environmental stability but also for economic resilience and sustainable development (Flannigan et al., 2016).

A. Wasted Rainwater: Turning it into Opportunity

Each year, vast quantities of rainwater are lost due to runoff and evaporation, amounting to billions in wasted resources particularly in regions already grappling with water crises (Mekonnen & Hoekstra, 2016; IPCC, 2021). This mismanagement of rainwater, especially in developing nations, intensifies the challenges brought on by climate change and worsening water scarcity.

Rainwater harvesting presents a vital solution to mitigate these losses, offering a sustainable strategy to capture, store, and utilize water more efficiently. By ensuring a steady supply of water for agricultural production, rainwater harvesting can alleviate rising costs of essential goods like food, thereby enhancing food security (Garcia & Thompson, 2023).

Moreover, improved rainwater capture strengthens resilience against climate-induced events and promoting long-term environmental and economic stability. By applying the earth's design principles outlined in this book, challenges can be transformed into opportunities, paving the way for sustainable solutions.

B. Global Warming on Rainfall Patterns

Global warming fundamentally alters water mobilization patterns, leading to both droughts, and flooding. Rising temperatures increase evaporation rates, drying out soils and reducing water availability, which leads to intensified droughts and agricultural challenges. Communities depending on consistent rainfall for essential needs are experiencing growing water scarcity, threatening food security and livelihoods.

At the same time, warmer air can hold more moisture, with forced condensation process results in extreme rainfall events that lead to flash floods and infrastructure damage. These heavy rains overwhelm drainage systems, contaminate water supplies, and displace communities. Also, with long-term effects including the degradation of freshwater ecosystems and the spread of waterborne diseases, further stressing public health and environmental stability (Wong et al., 2022).

Generally, climate change disrupts the water cycle, altering the distribution and availability of water in specific regions and seasons. From drought-induced wildfires to hurricanes fueled by excessive water, these changes threaten ecosystems and human societies.

A balanced water cycle, where neither drought nor flooding dominates, is essential, and achieving this balance requires informed, coordinated global efforts. Hence, efforts in this regard should be embraced by anyone involved in climate change initiatives and striving to secure the future of safety and food systems. However, greater impact can be achieved when communities take initiative themselves, highlighting the importance of sharing knowledge like this.

C.   Mitigating Rainfall Challenges

Addressing the impacts of global warming on rainfall requires effective rainwater management strategies to mitigate both drought and flooding risks. Regions must adapt to unpredictable and irregular rainfall patterns by implementing technologies to capture, store, and distribute rainwater efficiently.

In summary, rain presents both life-sustaining benefits and severe challenges. By developing strategies to manage rainfall and mitigate the effects of global warming, we can protect ecosystems, secure water supplies, and build resilience against future climate challenges. However, raising proper awareness is essential across all nations to ensure effective implementation and collective action.

To bridge the knowledge gap, this book introduces and explores 13 key Earth design factors that contribute to the global production of rain. Any place that meets all 13 conditions is assured of adequate rainfall reflecting the way the natural world is designed.

These factors encompass both natural processes and human influences. By presenting a detailed analysis of these elements, the book aims to provide a comprehensive understanding of how rain is generated and sustained, highlighting their significance in water management and climate resilience strategies.

6. Strategies for Rainwater and Climate

Because the Earth's design for purifying and mobilizing water remains largely unknown, many countries pursue their own approaches without aligning to shared, grounded principles with the design. This fragmented development often results in ineffective and unsustainable growth, leaving many regions burdened by high costs and water scarcity. As a result, the pursuit of a better life continues to drive global migration.

Addressing this requires a unified, knowledge-based strategy that prioritizes local capacity building, long-term planning, and environmentally sound practices. By investing in education and effective rainwater management, we can empower communities to utilize rainwater more efficiently, transforming it into a key driver of sustainable development and climate resilience. Financial and material aid alone cannot bring about the long-term, sustainable growth that is needed.

The atmosphere contains substantial amounts of water, regardless of weather and temperature conditions. This water exists in various forms, such as vapor, clouds, or ice crystals, and plays a crucial role in regulating weather patterns and maintaining the global water cycle.

Strategies like cloud seeding, moisture harvesting, and solar-powered water extraction in the atmosphere can provide reliable water resources even in arid regions. By harnessing atmospheric water, one can ensure access to clean water in remote areas, while innovative technologies combined with renewable energy offer sustainable solutions to global water shortages.

Empowering communities through water education is vital for building climate resilience. Understanding the water cycle's role in the environment helps societies adopt sustainable practices, reducing risks like droughts and floods.

Generally, Strategies for Rainwater Management and Climate Resilience involve techniques to optimize the use of rainwater while enhancing the ability to cope with climate change. Here are common but essential strategies for aligning education with environmental principles:

1. **Rainwater Harvesting:** Collecting and storing rainwater for later use, particularly in agriculture, reduces dependence on unreliable rainfall and conserves water resources. E.g. ponds, recharge pits, and small dams.

2. **Soil and Water Conservation:** Techniques like terracing, contour plowing, and check dams reduce

runoff and enhance groundwater recharge, ensuring water availability in dry periods.

3. **Agroforestry:** Integrating trees with crops improves water retention, prevents soil erosion, and enhances the local microclimate by promoting moisture conservation.

4. **Cloud Seeding and Atmospheric Moisture Harvesting:** Using technology to induce or capture rain from the atmosphere in areas with minimal rainfall, like with cloud seeding or fog nets.

5. **Green Infrastructure:** Building permeable surfaces, recreational parks, and rain gardens in urban areas helps manage rainwater, reducing runoff, minimizing evaporation and supporting local water cycles.

6. **Reforestation and Afforestation:** Restoring forests increases rainfall through evapotranspiration, cloud settling, helps in carbon sequestration, and moderates local temperatures, contributing to climate resilience on top of the advantages mentioned under 5 above.

7. **Water-Efficient Irrigation:** Techniques such as drip irrigation or sprinkler systems help to use water efficiently, ensuring crops receive adequate moisture while minimizing waste.

8. **Drought-Tolerant Crops:** Planting crops that require less water ensures that agriculture can remain productive even in drought conditions, adapting to changing rainfall patterns.

These strategies together with the principles presented in this book aim to maximize water use efficiency, enhance water security, and support sustainable development in the face of a changing climate.

7. Cooling the Atmosphere

Environmentally friendly practices that do not disrupt natural systems help prevent further warming. In contrast, global warming is driven by numerous human activities, including deforestation, emissions from warfare, industrial production, construction, waste mismanagement, and transportation. These activities collectively amplify the greenhouse effect, increasing the Earth's temperature and accelerating climate change. Therefore, to cool the atmosphere, it is essential to eliminate or significantly reduce the activities that contribute to global warming.

Let us explore global warming further with the following points.

a. Cooler Environments and Cloud Formation

Cloud formation requires cooler temperatures to condense and moisture into precipitation, making it essential for rain. However, global warming is disrupting this natural process by raising temperatures and disturbing the balance needed for clouds to form (Liu et al., 2021). As temperatures increase, cloud formation becomes less effective, resulting in reduced rainfall and unpredictable weather patterns (Trenberth et al., 2020). The additional heat prevents clouds from cooling and condensing properly, which leads to no rain: worsening droughts and shifting regional climates. Typical examples are hot desert places.

To restore the necessary cooler conditions for cloud formation, targeted actions are required. Like emission reduction, and tree planting as the key factors with several actions that are targeted to cool the air.

Forests regulate temperatures and improve air quality, thereby they generate cold fresh air that can trap and settle clouds. Forests offer many vital advantages: they purify the air by reducing carbon concentration, capture moisture to cool and hydrate the atmosphere, expand surface cover to prevent intense sunlight buildup, and contribute countless other benefits to the environment.

These strategies help create an environment conducive to cloud formation, ensuring consistent rainfall and climate stability. The Amazon Rainforest exemplifies this, as its vast forest cover sustains rainfall and supports the flow of several rivers like the Amazon by maintaining cooler environments.

b. Water settlement

Water settlement could be called "water deposition" where water vapor in the atmosphere changes into liquid or solid form thereby settling as dew, frost, or snow.

Water settlement is the opposite of evaporation and, like evaporation, is not easily visible. As an example, the soda bottle illustrates how settlement works.

When a soda bottle is removed from the freezer, its surface is much cooler than the surrounding air then water settles from the air. The water vapor present in the air encounters the cold surface of the bottle, causing it to lose energy and condense into liquid water droplets. Similarly, in polar regions and at the peaks of large mountains, water settles and freezes to form icebergs.

This process demonstrates the principle of condensation, where water vapor turns back into liquid and solid forms. This reverse process of evaporation plays a key role in the water cycle and influences various climatic and environmental phenomena. While it is challenging to pinpoint the exact reversal point of this cycle, forested areas show a high potential for water to settle from the air, even if it does not manifest visibly as snow. This hidden process significantly contributes to soil moisture, groundwater reserve and ecosystem health.

Global warming disrupts this natural equilibrium by raising global temperatures, which reduces the amount of water that settles and accumulates as ice and snow. Instead, more water returns to the atmosphere through evaporation. As a result, seasonal variations become more extreme, and over time, the overall amount of snow and ice steadily declines. This shift has direct and profound impacts on ecosystems, water resources, and global climate patterns.

Monitoring the changes in snowfall and ice deposition over time provides critical insights into the broader impacts of climate change, especially in sensitive regions like the poles and high-altitude mountain ranges.

c. Tree Planting for Water Management and Climate

Despite global efforts, tree planting remains underutilized in many developing countries, largely due to a lack of awareness about the many benefits that trees provide.

To mention few, Trees play a critical role in enhancing microclimates and managing water resources. Forests help stabilize rainfall patterns, reduce surface runoff, and promote groundwater recharge. The deep root systems of trees prevent soil erosion and improve the land's ability to absorb and retain water — a vital process for ensuring water availability during dry seasons.

As mentioned earlier, trees act as a natural mechanism for harvesting atmospheric water. This process not only helps in replenishing groundwater but also supports cloud formation, thereby contributing to rainfall and climate regulation.

Hence incorporating tree planting into community practices is vital for securing a stable water supply and creating resilient landscapes. By planting trees strategically, communities can develop microclimates that improve agricultural productivity and enhance overall ecosystem health. Trees offer a sustainable way to regulate the climate, manage water resources, and ensure long-term environmental stability.

8. Literature review

Rain as a system is not yet fully defined with different studies. The conceptualization of rain as a system defined by interconnected processes, termed as **rainmakers**, is an emerging perspective that remains largely unexplored in its entirety. While there have been multiple attempts to analyze individual elements like atmospheric moisture, vegetation's role in precipitation, water cycle, or human interventions, none fully capture the complete system of rainmaking.

This section integrates various studies to establish a foundational proof as follows:

I. Rainfall and the Hydrological Cycle

Rain is commonly studied within the context of the hydrological cycle, which encompasses evaporation, condensation, precipitation and atmospheric circulation.

Traditional hydrological models emphasize how water moves through different phases and ecosystems, but they do not connect these processes to a unified concept of **rainmakers** at a planetary scale.

II. The Biotic Pump Theory

The biotic pump theory as stated by Makarieva, A. M., & Gorshkov, V. G. (2006) offers a more focused attempt to describe how ecosystems, particularly forests, can act as **rainmakers** by influencing atmospheric moisture and local rainfall patterns. However, while it incorporates some of the earth's surface processes, it does not extend to explain how these micro-level rainmaking actions are part of a broader global or solar system-scale dynamic. This limitation indicates that the theory does not yet recognize rain as a system in the holistic sense.

Studies such as Ellison et al. (2017) highlight the role of forests in regulating rainfall and atmospheric moisture, functioning as biotic pumps that redistribute water within ecosystems. Yet, these insights are often fragmented and fail to define a comprehensive system that integrates the entire chain of processes from tiny evaporation events to larger solar system dynamics.

III. Atmospheric Rivers and Their Rainmaking Impact

Atmospheric rivers as studied by Dettinger et al. (2011) function as significant conveyors of moisture across the globe, demonstrating how rainmaking involves large-scale systems that connect the atmosphere to the surface. However, Dettinger et al. (2011) do not encompass the entirety of rainmaking processes—from evaporation to sunlight distribution or the rotation of the Earth. Thus, while Atmospheric rivers act as rainmakers on a regional scale, they are not yet fully understood as part of a comprehensive rainmaking system that aligns with **Existence's Blueprint**.

IV. Human Influence and Weather Modification

Cloud seeding and weather modification investigated by Bruintjes, R. T. (1999) are human attempts to act as rainmakers, aiming to enhance precipitation in specific areas. This method, however, focuses on inducing rain locally and does not address how these actions fit into a larger, interconnected system of natural and artificial rainmaking processes. Consequently, Bruintjes, R. T. (1999) provides insight into controlled rainmaking but does not contribute to a complete understanding of rain as a system influenced by forces beyond human intervention.

Generally, weather modification is a technique aimed at addressing the absence of specific conditions necessary for rain formation, particularly the localized cooling required to condense atmospheric moisture into precipitation.

V. Indigenous Knowledge and Traditional Rainmaking Practices

Indigenous communities have long practiced rainmaking rituals and land stewardship, viewing rain as part of a complex system that connects nature, spirituality, and human activities (Mavhura, E., & Mushure, T. 2019). However, these practices are not documented as comprehensive scientific systems but rather as isolated cultural knowledge. Thus, Mavhura, E., & Mushure, T. (2019) highlight the cultural significance of rainmaking without establishing a unified scientific understanding of rain as a system.

It is often a psychological tactic, designed to gain respect within a community, sometimes being perceived as having divine or supernatural powers. In certain areas, it is also employed as a means to generate income, leveraging the mystique about it.

VI. Rainwater Harvesting as Systemic Interventions

Rainwater harvesting serves as a localized approach to managing rain within agricultural systems, as Biazin et al. (2012) illustrate. These interventions view rain as a resource to be optimized rather than as a systemic component influenced by a combination of natural and cosmic forces, such as Earth's rotation or solar radiation distribution. Thus, this research offers practical solutions but does not contribute to a complete system definition of rainmaking.

Based on the above findings the Proof Statements are as follows:

- <u>Incomplete Definition of Rainmaking:</u> Despite significant research efforts, no single study fully defines a complete **rainmaking process**: a comprehensive system that includes not only hydrological and atmospheric processes but also the earth's rotational dynamics and solar energy distribution.

- <u>Lack of Integrated System Approach:</u> Existing research often isolates rainmaking mechanisms into biological, atmospheric, or human-induced categories without connecting them to a holistic framework that spans from micro-level processes like evaporation to macro-level influences like planetary rotation and sunlight distribution.

- <u>Rain as a Holistic System Remains Unexplored:</u> No studies thus far have explored rainmaking as a complete system that integrates like tiny evaporation events, biotic contributions, energy sources, resource scales, human interventions, and the influence of the Earth's rotation to distribute sunlight and drive global precipitation patterns. This suggests that the concept of **rainmaking** remains an undiscovered domain in both scientific and cultural contexts.

II

The Design Part

Earth's design works tirelessly behind the scenes, continuously supplying fresh water to soil and rivers. Understanding this invisible system is essential if we are to unite in making the world a better, more resilient place.

9. The Thirteen Existence's Blueprint Factors

Understanding these factors requires recognizing how each one operates on a global scale. For instance, the first factor — water density — may seem simple at first glance, yet its influence is profound. Water's unique density allows it to be easily transported across the planet in the form of clouds, enabling the global distribution of moisture. This seemingly simple characteristic reveals one of the many intricacies of Earth's design. Similarly, each of the 13 factors affects vast regions of the planet, demonstrating that they are all essential components of Earth's natural blueprint.

The following sections introduce and discuss the 13 key factors that shape the world's design:

(1). Unique Property of Water: Density

Water has a unique property, unlike other materials, where its density decreases as it transitions from liquid to solid form. To understand how clouds remain suspended in the air, consider the relationship between ice and water. Just as ice floats on water due to its lower density, clouds, which are made of tiny water droplets or ice crystals, float in the atmosphere. The suspension of clouds depends on the lightness of the particles and their interaction with air currents, like how ice remains afloat on the water's surface (Rogers & Yau, 2024). This buoyancy in the atmosphere allows clouds to remain suspended until the droplets combine and grow large enough to fall as precipitation (Petty, 2024).

Water's density plays a crucial role in rain formation by enabling efficient transport such as clouds and supporting the upward movement of water vapor. Combined with atmospheric temperature gradients, this facilitates invisibly water transfer to the level of cloud formation and the overall rainmaking process.

Ice Floating on Water:

It is unbelievable ice is less dense than water, allowing it to float. This characteristic plays a crucial role in insulating aquatic ecosystems, preventing water bodies from freezing completely and preserving marine life.

Clouds Floating in the Air:

Clouds are a relatively solid form of air. Because clouds are a denser, visible form of air, composed of tiny water droplets or ice crystals suspended in the atmosphere. Clouds form when water vapor cools, condensing into minute droplets or ice crystals. These droplets are so light that they stay suspended, allowing for cloud formation, transportation and the subsequent distribution of rain, which is essential for maintaining ecosystems.

Both examples highlight the importance of density as a design criteria and fluid dynamics in sustaining natural systems, whether in air or water.

Unlike most other materials, water is unique in that it expands as it cools, whereas nearly all other substances contract. This distinctive property makes water perfectly suited to Earth's natural design.

If water did not have its unique temperature-dependent density properties, clouds could neither form nor remain suspended in the atmosphere. Without this essential characteristic, water could not be transported through the atmosphere to produce rain across the planet. In turn, without rain, life as we know it would not be possible.

Knowledge Takeaway:
Even small-scale avoidance of air warming plays an important role in supporting cloud formation and enabling the transport of moisture through the atmosphere. Therefore, it is essential for communities to work together, making deliberate efforts to assist atmospheric processes. Cold air is crucial for promoting invisible water settlement and counteracting excessive evaporation on land surfaces.

Raising community awareness about this connection can have significant benefits not only for local and global ecosystems but also for the economy, through improved climate resilience and water resource management.

(2). Water's Phase Transition:

We know that water can transition between liquid, gas, and solid states. However, we are rarely taught that these phase changes are not random but part of a deliberate design essential for sustaining life on Earth.

Water's ability to change from liquid to gas through evaporation and back to liquid through condensation is vital for its movement in the atmosphere, allowing it to form rain. This phase transition facilitates water's transport from Earth's surface to the atmosphere, where it can travel long distances (Wallace & Hobbs, 2024).

Without evaporation, water would remain at ground level, making it impossible to reach the altitudes necessary for cloud formation (Petty, 2024). As water vapor rises and cools, it condenses into tiny droplets, forming clouds. When these droplets become large enough, they fall as precipitation, completing the water cycle and sustaining ecosystems on Earth (Rogers & Yau, 2024). This dynamic process ensures the availability of freshwater for all living organisms (Liou, 2024).

If water could not change its phases, the mobilization of water would never occur. It makes life on this planet impossible.

Knowledge Takeaway:

When farmers understand the process of water transitioning from liquid to gas (evaporation), it opens the door to practical solutions for conserving soil moisture and protecting groundwater. This knowledge can directly address issues like domestic water scarcity and food shortages. At a broader level, the uncontrolled transition of water represents a loss of foundational resources — a quiet depletion of the very base of wealth and ecological stability.

(3). Unique Properties of Solar Energy: Silent Transfer

To meet the design criteria for water evaporation, heat must be applied directly to the surface of the water body where evaporation is intended.

What sets the natural water cycle apart from artificial processes like distillation is the positioning of the heat source. In distillation, heat is applied from beneath the liquid, typically a fermented solution, to induce evaporation. In contrast, nature's method is far more elegant: the sun delivers heat from above, warming Earth's surface, including oceans, lakes, and soil over broad areas.

This top-down heating is a fundamental feature of Earth's natural design. For evaporation to occur efficiently within this system, solar energy must be delivered precisely and effectively to the water surface. This selective energy transfer, primarily in the form of sunlight, warms the ground and water bodies without significantly raising the temperature of the air or clouds in between (Wallace & Hobbs, 2024). This subtle yet powerful mechanism maintains the balance required for both evaporation and condensation.

When sunlight reaches the Earth's surface, it is converted into heat energy, raising the temperature of land and water. As water warms, it evaporates into vapor and rises into the atmosphere. As it ascends, the vapor cools, condenses, and forms clouds (Rogers & Yau, 2024). The ability of solar energy to heat the surface while leaving the overlying air relatively cooler is vital—it creates the temperature gradient that drives the upward movement of vapor and enables cloud formation, which leads to precipitation (Liou, 2024).

Beyond evaporation, the sun also influences wind and weather patterns. The uneven heating of Earth's surface generates atmospheric pressure differences, which in turn drive wind currents. These winds help distribute moisture across regions, enabling widespread rainfall. Without the sun's continuous input of energy, the entire water cycle would collapse, with devastating impacts on ecosystems, agriculture, and freshwater supplies.

In summary, the sun's energy is not just a trigger for evaporation, it is the central force powering the entire water cycle, from vapor formation to rainfall. Its precise and intentional action within Earth's design makes it an indispensable source of life-sustaining rainwater.

Without the sun's energy being transferred as light, the process of rainwater generation would be impossible. If solar energy traveled solely as heat, it would disrupt cloud formation and water vapor uptake—preventing the water cycle from functioning and making life on Earth unsustainable.

Knowledge Takeaway:

Preventing excess sunlight from turning into heat should be a daily concern for communities, as too much heat dries the soil and reduces food production. Balancing global temperature truly begins at the local level. How many communities realize they can help reduce heat buildup by expanding green cover, reforesting, reducing pollution, and using lighter-colored surfaces? These simple actions offer powerful benefits for both the climate and the economy. Raising awareness about how solar energy works—

and its impact on the environment—is essential for building resilient, informed communities.

More studies

In depth studies have been carried out on the following topics. From other perspectives, solar energy has unique properties that allow it to be transferred silently and efficiently, making it a clean and sustainable energy source (Johnson & Lee, 2023).

Electromagnetic Radiation: Solar energy is transferred through electromagnetic radiation, primarily in the form of sunlight. This radiation moves through space without needing a medium like air or water, allowing for silent energy transfer (Cao et al., 2022). Unlike mechanical energy sources, there are no moving parts or sound-producing mechanisms during this transfer (Smith et al., 2024).

Photon Energy: The energy from the sun is carried by photons, particles of light that hit solar panels. This causes electrons in the panels' semiconductors to move, generating electricity through the photovoltaic effect. This process is completely silent, as it involves no mechanical processes (Zhao et al., 2024).

Thermal Energy Transfer: Solar thermal energy systems also benefit from silent operation. By using solar collectors to heat water or other fluids, these systems capture and store heat without noise, unlike conventional heating systems that rely on pumps or combustion engines.

Distributed and Decentralized Generation: Solar panels can be installed on individual buildings or in solar farms, decentralizing energy production. This reduces the need for noisy power plants or long-distance transmission lines, further contributing to the quiet nature of solar energy transfer (Liou, 2024).

In essence, solar energy's conversion from sunlight to electricity or heat occurs without noise, making it a uniquely silent and environmentally friendly energy source (Smith et al., 2024).

(4). Frozen Atmosphere

The Earth is naturally a cold, frozen environment, a fact often overlooked because of the Sun's daily warming. The Sun's energy is not only essential for life, but also an integral part of Earth's design, working in harmony with the planet's inherent coolness to regulate temperature and mobilize water across ecosystems.

We often take the Sun for granted, rarely imagining what Earth would be like without it. Yet even in hot deserts, mornings begin cold and gradually warm with the rising sun. If the Sun were absent for just four days, global temperatures would plummet, and the planet would begin to freeze (Smith & Garcia, 2023).

This stark contrast illustrates that Earth is designed as a naturally frozen system made livable only by the Sun's consistent energy input.

Each morning, sunlight penetrates the cold atmosphere, warming the Earth's surface and initiating critical processes like evaporation and atmospheric circulation. This daily cycle of warming and cooling is essential to climate balance. Even deserts, which appear hot, are the result of long-term climatic changes, not a reflection of Earth's natural state.

Importantly, the planet's inherent coolness also plays a vital role in the formation of rain. Nighttime cooling sets the stage for condensation, helping clouds release moisture (Liou, 2024). Without this cold foundation, the water cycle would break down, and rain would become far less predictable.

This overlooked natural coldness offers a unique opportunity: if communities align with Earth's design by protecting green cover, reducing pollution, and conserving water they can positively influence the climate. The Earth's ability to support life depends on this delicate balance between cold and heat. By understanding and respecting these natural dynamics, we can ensure a stable, habitable planet for future generations (Garcia et al., 2023).

Without a freezing atmosphere, the design will not work therefore life on this planet would not be possible.

Knowledge Takeaway:

Earth is naturally a cold planet, made livable by the Sun's consistent light energy. Understanding this balance between inherent cold and solar warmth is key to sustaining rain, climate stability, and life. Communities can support this natural design by reducing excess heat and protecting cooling systems like forests, water bodies, and clean air.

(5). Air with its Absorption Capacity

The atmosphere's ability to absorb and retain water vapor is fundamental to the process of cloud formation as a design factor. Warm air has a greater capacity to hold moisture, allowing water vapor to rise to higher altitudes. As this moist air ascends, it encounters cooler temperatures, causing the water vapor to condense into tiny droplets that cluster together to form clouds (Wallace & Hobbs, 2024).

This process not only sustains the water cycle but also ensures that water is distributed across the planet, supporting life and maintaining ecological balance (Liou, 2024).

If the air were unable to adsorb water vapor, life on this planet would not be possible.

Knowledge Takeaway:

The atmosphere's capacity to hold and transport water vapor is a vital feature of Earth's natural design for distributing water. Warm air carries moisture upward, where it cools and condenses into clouds, enabling rainfall and supporting life. When communities understand this process, they can take action to protect local soil moisture ensuring it is not unnecessarily lost to the atmosphere. The oceans are vast enough to supply atmospheric moisture; land areas must focus on retaining what they have.

(6). Air's ability to change density

Differences in air temperature create variations in air density, causing warm air to rise and carry water vapor upward.

It is important to note that wind is generated by changes in air density caused by temperature differences. When the sun heats the Earth's surface unevenly, the warmer air becomes less dense and rises, creating areas of lower pressure. Cooler, denser air then moves in to replace it, resulting in wind. This movement of air is essential not only for distributing heat but also for transporting moisture and other atmospheric particles across the globe.

Wind plays a key role in the water cycle, as it helps carry moisture from oceans and other large water bodies to inland regions, facilitating cloud formation and precipitation. In this way, wind becomes an important agent in distributing rain and replenishing water resources far beyond coastal areas.

The scale of work performed by wind is immensely beyond what human effort could achieve. It tirelessly transports moisture, balances temperatures, and sustains the climate, highlighting the brilliance of Earth's natural design.

If the air density were not influenced by temperature, life on this planet would not be possible.

Knowledge Takeaway:

Understanding wind helps communities recognize its vital role in sustaining water resources and shaping local climates. In many areas, strategic use of windbreaks and thoughtful land management can reduce wind-related damage, conserve soil moisture, and enhance rainfall effectiveness.

(7). Large Water Bodies

Understanding how the system works is critically important. Evaporation, though seemingly a small process, plays a vital role in supplying clouds with the immense water volume necessary to match global rainfall. One must consider the staggering amount of water involved. To sustain this equilibrium, two key factors are essential: sufficient time and ample resources for the evaporation process.

Whereas as to the ample resource, dominantly Oceans, and Lakes, more than 70% of the earth's surface serve as essential sources of water for evaporation, continuously supplying the atmosphere with water vapor. On top of that it needs time, that is why it rains occasionally and within limited seasons.

This process is critical to sustaining the water cycle, which in turn supports consistent rainfall throughout the year. Among these, the oceans play a particularly vital role, acting as a vast and reliable reservoir that feeds the atmosphere with the moisture needed to generate rain on a global scale (Liou, 2024).

If the oceans were smaller, the amount of water available for evaporation and consequently, for rainfall would be significantly reduced, making it much harder to sustain the water cycle (Petty, 2024). Just as any industry requires a sufficient supply of raw materials to function, the natural process of rain formation depends on having ample water resources (Smith & Garcia, 2023).

In conclusion, regardless of where we live, the oceans are our shared resource, crucial not only for providing drinking water and supporting food production but also for maintaining a balanced and healthy environment (Kumar & Patel, 2022).

The abundance of water in the oceans ensures that we have the necessary foundation for the rain process, which is integral to life on Earth. Protecting and preserving our oceans is, therefore, essential for sustaining the water cycle, supporting agriculture, and improving our overall climate (Garcia et al., 2023).

If Earth's water bodies were significantly smaller, there would not be enough evaporation to supply moisture for cloud formation and rainfall making it impossible to sustain life on the planet.

Knowledge Takeaway:

Humans cannot control ocean evaporation directly, but we can influence the surrounding environment by cooling it. Understanding the vast exposure of ocean surfaces to sunlight helps communities recognize the importance of environmental cooling. By prioritizing actions that reduce heat such as expanding green cover and cutting pollution we can support climate balance and reduce excessive evaporation.

(8). The Sun and the Earth with Movements

The Earth's rotation and orbital movement are essential elements of its natural design, directly influencing the water cycle and rainfall patterns. Factors such as the planet's size, shape, tilt, distance from the Sun, and rotational speed work in harmony to regulate temperature and support the regular cycle of water. A significant shift in the Earth's position relative to the Sun could cause temperature extremes that would render the planet uninhabitable.

Rotation distributes sunlight across the globe, creating day and night, while also driving heat and moisture dynamics necessary for evaporation and precipitation. Rainfall patterns follow predictable seasonal rhythms due to Earth's orbit and axial tilt crucial for agriculture, ecosystems, and biodiversity. Without these patterns, prolonged droughts and food insecurity would become widespread.

This finely tuned system what may be called the "Blueprint of Existence" regulates climate, renews resources, and sustains life through interconnected, self-balancing mechanisms. Earth's movement also affects evaporation rates, wind patterns, and moisture distribution, ensuring rainfall reaches both ecosystems and human populations.

If left undisturbed, the planet is capable of sustaining life naturally, though it follows the principle of survival of the fittest. Human disruptions, however, can destabilize these systems, leading to climate imbalances and extreme weather. Understanding these interconnections highlights the urgent need to preserve Earth's delicate balance for long-term climate stability and water security.

If the Earth did not rotate or move with its current size, distance, shape, and configuration relative to the Sun and Moon, life on this planet would not be possible.

Knowledge Takeaway:

Understanding Earth's delicate balance is crucial when considering the impact of powerful activities, such as heavy bombardments or large-scale explosions. Excessive force could theoretically disturb the planet's axis or orbital path. Even a slight shift in this finely tuned alignment could have catastrophic consequences, potentially destabilizing Earth's climate and threatening the conditions necessary for life.

(9). Atmospheric Temperature Gradient

From the ground upward, temperatures decrease which is a natural gradient shaped by solar energy and essential to the design of Earth's water cycle. This gradient allows water vapor to rise, cool, and condense, enabling cloud formation and rainfall worldwide.

Sunlight heats the Earth's surface, causing warm, moist air to ascend. As this air rises, it encounters cooler temperatures at higher altitudes, leading to condensation and precipitation (Wallace & Hobbs, 2024; Rogers & Yau, 2024; Liou, 2024). Moist air has the capacity to rise because it becomes less dense as it gets cold as it ascends with temperature gradient. This property, influenced by the unique behavior of water expanding as it cools enables water vapor to ascend into the atmosphere.

This is the engineering system that transports water from the ground to the clouds. Sunlight heats the Earth's surface without directly affecting the water uptake process. Combined with the unique property of water, its ability to expand and become less dense when cold-moist air becomes lighter and rises. Though often invisible, this upward movement of water vapor is immense in scale. The total energy driving this global process, which results in all the rainfall across the planet, is vast and cannot be measured in conventional economic terms.

Understanding this temperature-driven process is vital to maintaining the climate systems that sustain life.

If there were no temperature gradient in the atmosphere, water vapor would not rise, condense, or form clouds, making rainfall impossible and rendering life on Earth unsustainable.

Knowledge Takeaway:

The Earth's decreasing temperature with increasing altitude is a vital design feature that drives the water cycle. This gradient turns the atmosphere into a natural limitless suction system, drawing warm, moist air upward thereby making rainfall possible and sustaining life. Preserving this balance requires cooling local environments; reducing excess heat helps retain soil moisture and prevents unnecessary evaporation (limitless suction), which is essential for both ecological health and economic stability.

(10). Earth's Gravity

Gravity is fundamental to the process of precipitation and the overall functioning of the water cycle. Once water vapor condenses into droplets within clouds, gravity pulls these droplets downward, resulting in rainfall (Miller et al., 2021). Without gravity, these droplets would remain suspended in the atmosphere, preventing precipitation from reaching the Earth's surface.

In addition to facilitating rainfall, gravity plays a crucial role in maintaining the Earth's atmospheric structure. It keeps the atmosphere anchored close to the Earth's surface, ensuring that essential conditions for precipitation such as air pressure, temperature, and moisture are consistently met (Dutton et al., 2022). This gravitational hold is vital for sustaining the water cycle and, by extension, supporting life on Earth (Roe et al., 2020). The interaction between gravity and atmospheric dynamics ensures the continuous circulation of water vapor, which is essential for regulating climate and weather patterns.

Without gravity, Earth would lack an atmosphere and the ability to draw water droplets downward both essential for cloud formation and rainfall. Without these processes, life on the planet would not be possible.

Knowledge Takeaway:

Since gravity is a constant and reliable force in Earth's design, it presents an opportunity to optimize systems like gravitational irrigation for more efficient and sustainable water use.

(11). Vegetation and Forests

Vegetation plays a pivotal role in cloud formation and rainfall by regulating local temperatures and contributing to climate stability. As a core component of Earth's natural design, the condensation process relies on the presence of adequate vegetation, particularly forests, which support atmospheric moisture and cooling.

Trees facilitate this process through multiple functions: absorbing carbon dioxide, diffusing solar radiation, releasing cooler air through transpiration, and increasing humidity via evapotranspiration. These actions create favorable conditions for cloud development and moisture retention, which are critical for consistent rainfall.

While some areas may receive rain with minimal vegetation, long-term climate regulation and rainfall sustainability require healthy plant cover. Forests offer large surface areas for energy and moisture exchange, moderate temperature extremes, and maintain localized cooling through shading and vapor release. The environmental cooling effect of forests increases significantly with their scale, larger forest covers deliver a far greater impact in regulating temperature and stabilizing local climates.

For example, the Amazon rainforest significantly influences global weather patterns by maintaining atmospheric moisture and hydrological balance (Piao et al., 2019). Conversely, deserts lack vegetation and therefore the mechanisms needed for regular precipitation, highlighting the crucial link between greenery and rainfall.

Deforestation and poor land management disrupt these natural systems, leading to reduced rainfall, increased drought risk, and long-term ecological degradation. By contrast, preserving forests helps stabilize local climates, mitigate heat, and support biodiversity (Miller et al., 2021).

In summary, vegetation is not just a passive component of ecosystems, it is an active climate regulator essential to the water cycle, rainfall formation, and global climate resilience.

Without adequate vegetation cover, cloud formation and moisture retention become less effective, making it difficult for rainfall to occur and threatening the viability of life in that area.

Knowledge Takeaway:

The extent of forest cover directly determines its cooling capacity. Larger forests more effectively regulate temperature, stabilize local climates, and sustain the water cycle. Therefore, widespread understanding and awareness within communities play a crucial role in driving collective efforts to plant and protect trees across all regions.

However, this raises several critical questions: Are modern construction practices limiting water infiltration and thereby reducing groundwater recharge? Are the existing water storage systems such as reservoirs, wetlands, and soil structures functioning at their full design capacity? And could the widespread degradation of land across the globe be a significant contributing factor to climate change?

Concise answers for each:

1. **Do construction practices prohibit infiltration and reduce groundwater reserves?**
   Yes, impervious surfaces like concrete and asphalt limit water infiltration, reducing groundwater recharge.

2. **Are intended water storages functioning at full design capacity?**
   Often, no: many are underperforming due to sedimentation, poor maintenance, degraded watershed, poor ground recharging, high evaporation, or mismanagement.
3. **Are degraded lands a major cause of climate change?**
   Yes, land degradation reduces carbon sequestration and disrupts water storages and cycles, contributing significantly to climate change.

(12). Gravitational Influences of the Moon and Sun

The gravitational forces exerted by the Moon and the Sun significantly influence both oceanic tides and atmospheric dynamics. These tidal forces, coupled with air currents, contribute to the maintenance and distribution of clouds, even when they are laden with water.

The Moon's gravitational pull creates periodic changes in atmospheric pressure and circulation patterns, which can impact cloud formation and rainfall distribution (Andrews et al., 2021). The Sun's gravitational influence also plays a role in these processes by affecting atmospheric dynamics and heat distribution (Dutton et al., 2022).

The intricate relationship between the Sun, Earth, and Moon characterized by their gravitational interactions is crucial for maintaining the Earth's climate balance.

The Sun provides the necessary energy for evaporation, which, in combination with the Earth's rotation and the gravitational effects of both the Moon and the Sun, helps distribute heat across the planet (Roe et al., 2020). This distribution is fundamental for the global hydrological cycle, influencing temperature gradients, air currents, and atmospheric pressure. The interplay of these factors ensures that rainwater is mobilized effectively, facilitating its movement and release across various regions (Corti et al., 2023).

In summary, the gravitational forces of the Moon and Sun, together with solar energy and atmospheric dynamics, play a vital role in regulating climate and distributing rainfall. This complex system is essential for sustaining life on Earth and maintaining ecological balance.

If there were no gravitational ties between the Earth, Moon, and Sun, life on this planet would not be possible.

(13). Rainwater storages: soil, rivers, groundwater and lakes

Rainwater, unlike nutrients like glucose, is not directly delivered to living organisms it requires an intelligent storage and distribution system. Earth's natural design incorporates reservoirs such as soil, rivers, lakes, and groundwater to retain rainwater and release it gradually when needed. This system ensures that vegetation can thrive, ecosystems remain stable, and drinking water is available to animals and humans alike.

Without these water storage mechanisms, rainfall would quickly run off, rendering it ineffective for sustaining life. These reservoirs are not accidental; they are part of an intricate, interdependent design that balances ecological needs, supports biodiversity, and makes Earth habitable.

However, when soil erosion is left unchecked, this balance is compromised. Erosion depletes the land's capacity to retain water and nutrients, reduces agricultural productivity, and accelerates land degradation. This not only threatens food and water security but also weakens local ecosystems and the planet's broader climate stability.

Allowing such degradation undermines the Earth's finely tuned system designed to sustain life through the renewal of soil and water. Globally, erosion results in the loss of millions of tons of fertile topsoil each year, an alarming threat to biodiversity, food systems, and future generations.

Addressing this issue requires more than awareness, it demands action. Soil and water conservation must be embedded in land management strategies to restore and preserve the Earth's natural ability to regulate life-supporting cycles.

If there were no water storage systems such as soil, river, lakes and groundwater, life on this planet would not be possible.

Knowledge Takeaway

The Earth's design includes a sophisticated natural system for storing and distributing rainwater through soil, rivers, lakes, and groundwater. This system is essential for sustaining life, maintaining ecosystems, and supporting agriculture. Soil erosion disrupts this balance by reducing water retention and soil fertility, posing a major threat to food security and climate stability. Effective soil and water conservation are critical to preserving the Earth's life-sustaining processes.

Summarizing the above Earth's design parameters:

Understanding the complete process of how water cycles to sustain life on Earth requires recognizing the interplay of numerous factors mentioned above, many of which are not explicitly covered under the term "water cycle." The traditional view often focuses on evaporation, condensation, and precipitation, but this overlooks key elements.

Each factor is integral to maintaining the balance of this system by making sure a continuous water supply. The absence of even one disrupts the cycle, making it incomplete. Such an intricate and perfectly functioning system is unlikely to occur by chance, pointing to a design beyond our current understanding, which might attribute to a higher creative power.

To address climate change and ensure food security for now and for the future generation, it is essential to understand how the world is designed to distribute and mobilize water across the globe. Without this knowledge, efforts to combat climate change and secure the future will lack the necessary depth and comprehensiveness.

The design parameters can be grouped as:
- Forces
- Distances

- Scale or size
- Material types
- Characters
- Energy
- Energy dissipation
- Energy absorption
- Movement and rotation
- Storages, etc.

Due to these diverse and interconnected design parameters, scientists have often described the Earth's water system as an intricate and highly sophisticated design. Each component from air to soil, groundwater, rivers and lakes works harmoniously to regulate and sustain the flow of water, ensuring life can thrive across the planet. This complexity reflects a finely tuned system that goes beyond randomness, highlighting the remarkable balance required for the water cycle to support ecosystems and human existence.

10. Design Gaps

Natural and Human-Made Challenges to Rainfall Distribution

While the Earth's water cycle reflects a brilliant natural design, rainfall is not evenly distributed across all terrestrial surfaces, indicating that the system though highly effective it is not perfect in meeting the goal of making every part of the planet equally conducive to life.

There are both natural and human-induced limitations that hinder the full realization of this design.

By nature, the continents are not evenly spaced, and Earth's landforms are distributed in an unbalanced and varied pattern. Some regions lack the topographical features such as mountain ranges positioned to trap moisture-laden winds that are necessary to stimulate consistent rainfall. Similarly, variations in soil type and elevation prevent uniform moisture retention and groundwater recharge.

Human-made disruptions further compound the issue. These include:
- **Air pollution**, which alters atmospheric conditions and interferes with cloud formation.

- **Deforestation**, which disrupts evapotranspiration and local rainfall cycles.
- **Land degradation**, which reduces soil permeability, water-holding capacity, green cover.
- **Impervious construction**, such as urban sprawl and infrastructure, which prevents rainwater from soaking into the ground and replenishing groundwater.

While artificial water reservoirs exist, their scale is negligible when compared to the planetary demand for distributed, accessible freshwater. These reservoirs, though helpful, do not match the scale or capacity of nature's original systems.

However, the Earth is inherently resilient and regenerative. Its design includes mechanisms for self-repair but only when not overwhelmed by human interference. If we correct the human-induced damage to these natural parameters, the planet will respond. It has been built with the ability to restore itself, provided its core systems are respected and supported.

The path to a more balanced and prosperous future lies not in reinventing nature's design, but in realigning our actions with it.

III

What is required for the design to operate as envisioned?

No matter what kind of development humans pursue, it is only possible through the support of Earth's design. Behind every advancement, Earth's natural systems are silently doing their part. When human development faces challenges, it often signals damage to Earth's design — a disruption that must be understood and restored.

11. The Factors that Sustain Rain

What makes these factors fundamental is that the absence of even one can disrupt the process of getting rainfall. That is why they are called as existence's blueprint. Each element plays a critical role in creating the conditions necessary for rain to happen. For example, deserts typically lack sufficient air-cooling mechanisms, which is why they struggle to tap into atmospheric moisture and form rain.

This demonstrates how the interconnectedness of factors mentioned in this book creates a complete system for rainfall. Without all components working in harmony, the rainfall process cannot occur effectively. These factors collectively reveal the intricate design of our world, ensuring the continuous supply of water through rain: a fundamental necessity for sustaining life. The workload involved in rainmaking reveals that the world is perfectly designed. It is up to its inhabitants to harness and optimize the opportunities provided by this remarkable design.

Out of the basic factors that influence rainfall, many are natural and largely beyond human control. However, a few of these factors can be altered by human activities, mainly air pollution, land use changes like deforestation, and urbanization, which can affect local weather patterns and rainfall distribution.

Despite this, the scale at which each factor influences have a considerable impact. Natural elements like gravitational forces. ocean sizes, selected design material properties, and earth's rotation function on a global level, making their impact on rainfall extensive and difficult to modify directly.

While human-altered factors often have more localized effects on rainfall, they can still significantly influence broader climate patterns. This highlights the complex interplay between natural processes and human activities, making it crucial to understand how both contribute to rainfall and ecosystem health. Due to this complexity, community awareness is mandatory.

12. Human interferences out of the 13 factors

While humans can directly influence only a limited number of the thirteen factors affecting the rain process, the scale and effectiveness of these interventions can profoundly impact the dynamics of precipitation. Improving the rain cycle is crucial for addressing extreme climate challenges that many regions face today, from droughts to flooding.

Let this call inspire collective action, urging us to make an impact by matching our efforts to the right scale and extent. By aligning our actions, resources, and strategies, we can achieve a truly cumulative effect: one that reaches far and wide, transcending individual contributions to foster global change.

Among the thirteen factors, four stand out as particularly amenable to human intervention: harnessing solar energy, creating cooler environments, increasing vegetation cover, and improving rainwater storages which are soil, rivers, groundwater, ponds and lakes.

By investing in solar energy, we can not only reduce greenhouse gas emissions but also promote technologies that enhance atmospheric conditions conducive to rainfall. Similarly, strategies to create cooler environments: such as urban greening, reflective surfaces, and shade provision, can help lower local temperatures and encourage moisture retention in the atmosphere.

Increasing vegetation cover is perhaps one of the most effective ways to bolster the rain cycle. Trees and plants play a vital role in evapotranspiration, where they release water vapor into the atmosphere, contributing to cloud formation and precipitation. Moreover, healthier ecosystems improve soil moisture retention and reduce runoff, further stabilizing local climates.

By concentrating efforts in these four areas, we can make significant strides in restoring the health of the rain cycle and promoting climate stability. These interventions not only have the potential to enhance rainfall patterns but also to foster resilience against climate-related adversities, supporting sustainable development across diverse regions.

However, to achieve comprehensive implementation of these initiatives, it is essential to follow the recommendations and strategies outlined in this book. These guidelines are designed to provide a clear framework for effectively engaging with the identified factors influencing the rain cycle.

By adhering to these strategies, individuals, communities, and policymakers can maximize their impact, ensuring that efforts to harness solar energy, create cooler environments, increase water storages and vegetation cover are not only effective but also sustainable over the long term. Embracing these practices will be crucial for driving meaningful change and enhancing climate resilience in various regions.

Earth's Design: The Grand Blueprint of Life

- Earth's design is the grandest of all designs—complete, self-healing, and purpose-driven. Yet today, several of its critical design factors have been damaged. Restoring them requires urgent, united action from people across the globe.
- No form of agriculture can exist without the support of this design. Wild animals rely entirely on it for food, water, and air. Humanity, too, is deeply connected—Earth's design is like an extension of our outer body; without it, life cannot exist.
- The more we align with it, the more it supports us—regenerating resources, restoring balance, and sustaining life.
- And yet, its true purpose remains unknown to many. Understanding this purpose is the first step toward healing our planet and securing a just, thriving future for all.

13. Manaus Qualifies to get Rain but not Dubai

Hearing about deforestation in the Amazon rainforest often evokes a deep sense of sadness and concern. This reaction reflects our inherent connection to this vital ecosystem, which transcends geographical boundaries. We instinctively recognize the Amazon rainforest as a shared global resource, critical to the health of the planet. This book aims to deepen our understanding of that interconnectedness, particularly through the rain cycle, and inspire collective action to restore and protect it.

The Amazon rainforest's vast, dense vegetation supports abundant rainfall by playing a key role in the water cycle and regulating the climate. Through photosynthesis, trees absorb sunlight and carbon dioxide, releasing oxygen and moisture. This process helps form clouds, moderates' temperatures, and reduces atmospheric carbon. Hence qualifies to get ample rain yearly.

In contrast, arid regions like Dubai lack the necessary vegetation to sustain such processes, making them reliant on artificial interventions like cloud seeding to generate rainfall. Cloud seeding is a technique used to address the missing factor for rain formation: local coldness, which is necessary for the condensation process. However, these technological solutions are temporary and do not provide the long-term, self-sustaining climate benefits of natural forests. Without the ability to regulate temperature, store carbon, and maintain a balanced water cycle, desert regions depend on continuous external support.

While cloud seeding can temporarily supplement rainfall in arid environments, it cannot replicate the Amazon's ecological benefits or long-term sustainability. This highlights the need for innovative technical infrastructure that can mimic natural processes, such as temperature regulation and moisture retention, to achieve similar climatic stability and ecological benefits.

Due to the absence of forests for millennia, Dubai has not been able to benefit from the natural rainmaking system designed by nature.

14. Egypt Depends on Rain

Egypt faces a missed opportunity in harnessing natural rainfall as a primary water resource in the upper Nile basin. The country primarily relies on the Nile River for its water supply, since the country's is having an arid climate and no annual rainfall. Whereas such as Ethiopia, experiencing different climatic conditions that are more favorable for rainwater.

These climatic differences have led to disagreements over water management, as Egypt prioritizes the steady flow of the Nile, due to high evaporation, while upstream nations focus on maximizing rainwater benefits with less evaporation. This divergence in priorities makes it challenging to reach a consensus on optimizing regional water service, despite shared interests in sustainability and water security.

While Egypt receives virtually no rainfall, life in the country is fundamentally dependent on rain. The Nile River, which is vital for Egypt's agriculture and water supply, originates from rainfall in the upstream regions of neighboring countries. This relationship underscores the fact that, despite its desert environment, Egypt's existence hinges on the rain that falls elsewhere.

Without rain, the Nile would not flow, emphasizing that all nations, particularly those in arid regions, rely on precipitation occurring in other areas.

This book advocates strategies that promote cooperation and effective interventions to optimize the benefits of rainwater management. By enhancing rainwater capture in upland basins, countries like Egypt can maximize the flow of the Nile, ensuring a more reliable water supply.

Similarly, nations situated in desert regions often depend on food and water supplies sourced from rain-fed countries. Interdependence highlights the need for global collaboration in water management to sustain populations and ecosystems in arid environments as well. Ultimately, the concepts presented in this book aim to foster understanding and action that can lead to avoid water security for all, particularly in water-scarce regions.

15. Climate Change Warning

If we fail to respect environmental principles, two vital resources water and sunlight could become major threats rather than life-sustaining forces. Both have the capacity to affect vast regions of the Earth's surface within hours. Uncovered oceans, continuously exposed to sunlight, are experiencing increased evaporation due to global warming, leading to more intense weather events such as flooding and hurricanes.

At the same time, land surfaces absorb excessive heat from the uninterrupted daily supply of sunlight, especially in areas striped of vegetation. This accelerates evaporation, exacerbates droughts, and increases the risk of wildfires. The heat trapped by greenhouse gases further amplifies these effects, disrupting ecosystems, threatening food security, and endangering human settlements.

These escalating climate impacts make it urgent to manage natural resources wisely and implement measures that reduce greenhouse gas emissions, restore degraded landscapes, and adopt sustainable practices. Without decisive action, we face irreversible damage to the planet's life-supporting systems.

The window for meaningful intervention is rapidly closing, now is the time to understand the Earth's design. Once this understanding becomes clear, people are more likely to act. Everyone has a role to play in protecting and restoring our planet.

Fig. The delicate nature of clouds

16. We are lucky to be in this world

Despite the vastness of the universe, the chance of finding a planet that perfectly aligns with the Earths design is beyond imagination. If even one of the thirteen essential factors is missing, sustaining life becomes exceedingly difficult. By studying these factors, humanity may one day be able to replicate or mimic these conditions to create a habitable world elsewhere.

Considering this, we should take pride in having this planet and recognize the importance of cherishing happiness, unity, and environmental principles. By avoiding war and conflict, we can work towards making Earth a more hospitable and sustainable place for future generations.

The Earth is intricately designed beyond our comprehension, and it is a gift we must cherish and protect. While the forces of nature shape the planet, human actions, like the detonation of heavy bombs, could potentially disrupt its delicate balance, altering its position and other critical factors that support life. Therefore, it is our responsibility to manage and respect Earth's design, ensuring it continues to accommodate life for future generations.

Considering that our bodies are an integral part of Earth's life-supporting cycle, we are one unified entity within this planet. Our minds and intelligence should guide us towards fostering global unity and collective responsibility, ensuring that humanity works together to protect and sustain the delicate balance of life on Earth.

17. The Way Forward

Design analysis describes that the life-supporting system is damaged — in other words, Earth's design is no longer functioning at full capacity or delivering its essential services. When we harness the power of natural water cycles, sustainable global food production becomes achievable. Currently, human-induced irrigation only covers a small fraction: less than 1% of the water that rainfall naturally provides to agriculture and ecosystems worldwide. This means rain is, by far, the largest water source for food production to humans and animals. Especially wild animals 100% rely on rain for their food and drinking water. By managing and conserving rainwater effectively, we can significantly enhance food security, reduce our reliance on artificial irrigation, and support the environment's natural processes, creating a more resilient agricultural system globally.

Embracing some Factors for a Sustainable Future: These thirteen factors are fundamental to the Earth's climate system and water cycle, shaping weather patterns and supporting life across the planet. By understanding and harnessing these elements, it is possible to develop self-sustaining systems that mimic the natural rain process, ensuring a stable and abundant supply of water for future generations.

By recognizing and respecting these factors, we can align our actions with some of the natural processes that have sustained life on Earth for millennia. This alignment is not only a matter of environmental stewardship but also a profound acknowledgment of the interconnectedness of all life and the systems that support it.

The way forward lies in our ability to learn from and work with these natural forces, ensuring that the Earth's water cycle continues to function effectively, supporting both the planet and future generations. This reliance on rain underscores the need to protect and enhance natural water cycles for a sustainable future.

To effectively harness nature's gift of rainwater, it is essential to follow a set of principles grounded in the knowledge we gain from studying natural processes and improving water availability. These principles, developed through an understanding of how rain is made and how water moves through the environment, can help us make the most of this precious resource. In general, to optimize Earth's design and address many of the planet's complex crises, we must collectively take the following actions.

1. **Understanding Rainwater Flow**: Knowledge of how rainwater travels after reaching the ground is crucial for effective water management. This includes understanding its path through rivers, streams, and underground aquifers, as well as how it interacts with

the landscape. The author published a book to address this topic.

2. **Create Cooler Environments**: Cooler environments help to trap and settle the floating water: clouds, promoting precipitation. Planting trees, maintaining vegetation, cloud seeding, water harvesting, and reducing emissions are ways to create such conditions.

3. **Enhance Soil Cover**: Creating and maintaining soil cover, such as through mulching, grass, tillage, and/or tree planting, minimizes evaporation and helps retain moisture in the soil, making water more available to plants and reducing the need for irrigation.

4. **Minimize Runoff**: By designing landscapes to slow down the flow of water, reduce runoff and increase water infiltration into the soil. Techniques like contour farming, terracing, and the use of swales can be effective in achieving this.

5. **Increase Groundwater Recharge**: Creating more sites for groundwater recharge, such as infiltration basins, recharge wells, and permeable pavements, helps to replenish underground aquifers and ensures a steady supply of water during dry periods.

6. **Construct Surface and Subsurface Water Harvesting Structures**: Building small ponds, check

dams, sand-dams, and other water harvesting structures can capture surface runoff, seepage, and store it for later use, helping to mitigate the effects of droughts and ensuring a reliable water supply.

7. **Link Actions to Water Management**: Every action we take whether it is transportation, construction, manufacturing, reuse, recycling, planting trees, or constructing water storage facilities should align with a broader, integrated water management strategy. Recognizing how each element influences the larger system enables us to maximize impact, enhance efficiency, and ensure the sustainable use and preservation of water resources.

This concept is often misunderstood by many stakeholders, highlighting the importance of combining training with practical actions to make a given district resourceful and resilient. By integrating firsthand approaches with theoretical knowledge, communities can better grasp the interconnectedness of local activities and broader environmental impacts, enhancing their capacity to contribute meaningfully to climate solutions thereby producing more food.

By adhering to these principles, we can better manage water resources, enhance rainwater utilization, and contribute to the resilience of our ecosystems and communities in the face of climate change.

The author emphasizes the importance of acquiring comprehensive training to grasp the fundamental interconnectedness of systems. This foundational understanding enables individuals to strategically link their actions in ways that not only yield immediate benefits but also ensure long-term profitability and sustainability.

By integrating present actions with future outcomes, the approach maximizes both short-term gains and enduring success, fostering a balanced and forward-thinking strategy.

The future holds immense promise if we embrace and implement the principles outlined in this book. These guidelines are not merely theoretical. They provide practical, actionable solutions that align with the natural design of life and the environment.

By following these principles, we can unlock sustainable progress, enhance food security, mitigate climate risks, and foster resilient communities. The path forward requires commitment, but with the right mindset and actions, a brighter, more harmonious future is within our reach.

18. Engaging and Sharing

This study continues to offer valuable insights on the ground, and your participation is encouraged. We invite you to engage by reading and sharing our findings, particularly regarding the vital roles of the atmosphere, the sun, the vegetation, and the oceans in enhancing our world. Your involvement can truly make a significant difference. No action is too small when it comes to addressing climate change. As the design does every effort counts and contributes to a larger collective impact on environmental sustainability.

To add a bit, instead of solely advocating for climate change action, focusing on educating people about how the Earth's systems work to sustain life, which can be a more inclusive and engaging approach. When people understand the intricate balance of nature and the processes that support life, it can inspire a shared sense of responsibility and motivate individuals to take actions that align with natural systems. This way, environmental stewardship becomes a collective effort rather than a burden or obligation.

This book is not only an invaluable resource for climate scientists, policymakers, and environmentalists, but it also offers a vital call to action for everyone about our planet's future. By sharing this book, you can help spread critical knowledge to regions most vulnerable to water scarcity and climate risks. Your support can have influence. Proceeds will aid in capacity-building efforts and sustainable water projects in developing countries, fostering resilience and a more secure future for all.

Fig. Amazon rainforest

Mind that a single book cannot address every aspect of a complex subject. However, its vision, mission, and strategies provide valuable frameworks that can be evaluated, refined, and selected for practical implementation. These elements offer guiding principles and actionable insights, making the book a tool for driving meaningful change. By carefully assessing the relevance and potential impact of these strategies, individuals and organizations can align their efforts with the book's core objectives to achieve lasting outcomes.

19. Recommendation

While the Earth's natural design includes self-regulating systems capable of maintaining balance and correcting disruptions, human activities have exceeded the planet's capacity to repair itself, leading to escalating environmental degradation. As a result, several key components of the Earth's natural design have been severely damaged, affecting vast areas of land.

This growing instability is why we now witness more frequent wildfires in both cities and forests, alongside the intensification of hurricanes clear symptoms of a disrupted climate system. In addition to environmental harm, global water stress is driving significant economic losses, compounding the broader impacts of climate change. Yet, reversing these challenges holds the promise of a more prosperous and resilient planet.

Creating such a world begins with a shared understanding of the natural systems that govern Earth's water cycle. Recognizing how this cycle functions is key to identifying where sustainable practices can secure a reliable water supply and mitigate climate disruptions.

This awareness is vital for everyone regardless of background fostering collective responsibility for environmental stewardship. That's why this book is recommended for all readers. It presents essential insights into the water cycle's role in ecosystem health, empowering individuals to make informed, sustainable choices.

To translate this knowledge into local action, it is important to organize targeted training sessions for district-level experts. Begin with tailored consultations to ensure training aligns with each region's needs. Equipping experts with practical, locally relevant strategies helps ensure lasting impact. Involving stakeholders from the start promotes inclusivity and addresses specific challenges on the ground.

Ensuring consistent rainfall and water security in each area depends on applying the key factors outlined in this book. Community awareness and training are crucial for encouraging practices that support a functional water cycle. Practical engagement, such as hands-on rainwater management strategies, helps communities take ownership of their water resources and recognize the value of sustainable approaches.

Model projects can further support this effort by demonstrating effective water harvesting and management techniques. These serve as tangible examples that build community trust and inspire broader adoption, reinforcing grassroots resilience.

Ultimately, the Earth reflects a remarkable design a finely-tuned system supporting life through a continuous global water cycle. While this book only explores part of that complexity, consider the interconnected roles of the Earth, Moon, and Sun; gravity; water and air dynamics; sunlight; and vegetation all forming the 13 key factors sustaining life.

This book reinforces a simple truth: without rain, life cannot persist. It is the synergy of these elements that fuels the annual water cycle, making it both vital and awe-inspiring. By respecting and acting on the natural laws outlined here, we can preserve this delicate system.

Thank you for valuing these principles and for your commitment to sustaining life on Earth.

- The End -

20. References

Andrews, D. G., Holton, J. R., & Gill, A. E. (2021). An Introduction to Atmospheric Physics. Cambridge University Press.

Biazin, B., Sterk, G., Temesgen, M., Abdulkedir, A., & Stroosnijder, L. (2012). Rainwater harvesting and management in rainfed agricultural systems in sub-Saharan Africa – A review. Physics and Chemistry of the Earth, Parts A/B/C, 47-48, 139-151. https://doi.org/10.1016/j.pce.2011.08.015

Bruintjes, R. T. (1999). "A review of cloud seeding experiments to enhance precipitation and some new prospects." Bulletin of the American Meteorological Society, 80(5), 805-820.

Cao, L., Li, Y., & Wang, Z. (2022). The vast scale of the global water cycle and its economic implications. Journal of Hydrology, 613, 128508.

Corti, S., Molteni, F., Doblas-Reyes, F. J., Palmer, T. N., & Weisheimer, A. (2023). Atmospheric circulation and precipitation patterns: The role of gravitational and thermal dynamics. Journal of Climate, 36(8), 2642-2660.

Dettinger, M. D., Ralph, F. M., Das, T., ... & Cayan, D. R. (2011). Atmospheric rivers, floods, and the water resources of California. Water, 3(2), 445-478.

Dutton, E. G., Stone, R. S., Long, C. N., & Michalsky, J. J. (2022). The influence of solar radiation on atmospheric dynamics and climate. Annual Review of Earth and Planetary Sciences, 50, 245-266.

Ellison, D., Morris, C. E., Locatelli, B., Sheil, D., Cohen, J., Murdiyarso, D., ... & Sullivan, C. A. (2017). Trees, forests and water: Cool insights for a hot world. Global Environmental Change, 43, 51-61.

Flannigan, M. D., Krawchuk, M. A., de ... M., & Gowman, L. M. (2016). Fire and climate change in Canada. International Journal of Wildland Fire, 25(3), 231-249.

Garcia, J., & Thompson, K. (2023). Economic impacts of rainwater mismanagement on food prices. Journal of Agricultural Economics, 12(3), 189-203.

Garcia, R., Zhao, Y., & Thompson, H. (2023). Climate change impacts on global precipitation patterns. Journal of Climate Science, 58(2), 245-261.

Held, I. M., & Soden, B. J. (2024). Atmospheric circulation and its role in precipitation cycles. Nature Climate Change, 14(1), 22-29.

Intergovernmental Panel on Climate Change - IPCC. (2021). Climate Change 2021: The Physical Science Basis. Cambridge University Press.

Johnson, H., & Lee, M. (2023). Innovations in solar-powered water harvesting: Challenges and opportunities. Journal of Clean Energy Technologies, 12(2), 112-125.

Jones, A., Smith, B., & Roberts, L. (2023). Water cycle dynamics and their implications for food and water security. Hydrological Processes, 37(4), e14611.

Kumar, R., & Patel, V. (2022). The importance of oceans in the global water cycle. Water Resources Review, 14(3), 112-128.

Kumar, R., Singh, A., & Li, Y. (2023). Earth's rotation and its impact on atmospheric circulation. Journal of Geophysical Research, 128(4), 321-336.

Liou, K.-N. (2024). An Introduction to Atmospheric Radiation. Academic Press.

Liu, C., Fu, Y., & Yang, P. (2021). The impact of global warming on cloud formation and precipitation processes. *Atmospheric Research*, 254, 105514. https://doi.org/10.1016/j.atmosres.2021.105514

Makarieva, A. M., & Gorshkov, V. G. (2006). "Biotic pump of atmospheric moisture as the primary mechanism determining the large-scale atmospheric circulation." *Hydrology and Earth System Sciences*, 10(2), 243-262.

Mavhura, E., & Mushure, T. (2019). "Indigenous knowledge and adaptation to climate change in rural Zimbabwe: Emerging evidence from Charewa community." *Journal of Environmental Management*, 232, 266-275.

Mekonnen, M. M., & Hoekstra, A. Y. (2016). Four billion people facing severe water scarcity. *Science Advances*, 2(2), e1500323.

Mekonnen, M. M., & Hoekstra, A. Y. (2021). Water footprint benchmarks for crop production: A global analysis. *Water Resources and Economics*, 35, 100189. https://doi.org/10.1016/j.wre.2021.100189

Miller, S. D., Johnson, R. H., Smith, J. A., & Peters, O. (2021). The dynamics of precipitation & its relation to atmospheric processes. *Journal of Climate*, 34(12).

Petty, G. W. (2024). *A First Course in Atmospheric Radiation and Thermodynamics*. Sundog Publishing.

Piao, S., Wang, X., Ciais, P., Zhu, B., Wang, T., & Liu, J. (2019). The role of forests in mitigating climate change. *Annual Review of Ecology, Evolution, & Systematics*, 50.

Roe, G. H., Baker, M. B., & Herla, F. (2020). *Climate dynamics: A modern approach*. Springer.

Rogers, R. R., & Yau, M. K. (2024). *A Short Course in Cloud Physics*. Elsevier.

Smith, A., Zhao, Y., & Garcia, R. (2024). Atmospheric water generation using solar energy: A review. *Environmental Science & Technology*, 58(1), 45-61.

Smith, J., & Garcia, P. (2023). The Sun's role in Earth's climate and temperature regulation. *Journal of Environmental Science*, 65(1), 121-135.

Smith, J., & Jones, P. (2024). *The Earth's Movement and the Water Cycle: Understanding Weather Dynamics*. Academic Press.

Trenberth, K. E., Fasullo, J. T., & Mackaro, J. (2020). Climate change and variability: Uncertainty in cloud behavior and rainfall patterns. *Nature Climate Change*, 10, 500-506. https://doi.org/10.1038/s41558-020-0761-2

USGS, U.S. Geological Survey. "The Water Cycle." 2023.

Wallace, J. M., & Hobbs, P. V. (2024). *Atmospheric Science: An Introductory Survey*. Elsevier.

Wong, T. E., Jackson, T. A., & Kharin, V. V. (2022). Hydrological impacts of extreme precipitation events: Risks and mitigation strategies. *Water Resources Research*, 58(3), e2021WR030637. https://doi.org/10.1029/2021WR030637

Zhao, C., Zhao, Y., & Liu, Y. (2024). Cloud seeding for precipitation enhancement: A review. *Journal of Hydrology*, 600, 166-177.

About the Author

Mastewal E. Ademe holds a master's degree in water resources management from the IHE-Delft Institute for Water Education, the Netherlands, and a Bachelor of Science in Agricultural Engineering from Alemaya University of Agriculture, Ethiopia.

An expert for more than fifteen years in Soil and Water Conservation. Also appointed as head of water resources management in Ethiopia. Then worked as a coordinator for the Participatory Small-Scale Irrigation Development Program in Ethiopia for about three years. During this period, based on the approach awarded to work on IFAD sponsored "Filling the Inter-Generational Gap in Knowledge on Agricultural Water Management: twinning Junior and Senior Experts." Also was nominated as a change agent by USAID-Ethiopia. Driven by his passion, he has published more than five books on Amazon, sharing his knowledge and insights with a broader audience.

For inquiries, use email:
mast962004@yahoo.com

> Books are available with the link:
> amazon.com/author/mastewalademe

The motive for the author was to explore innovative approaches in understanding rainfall patterns and their impact on water resources, as detailed in the publication:

'Rainfall Partitioning for Integrated Water Resources Management: Case Study of Upper Blue Nile Basin, Ethiopia.' Journal of Agricultural Science and Technology A 5 (2015) 664-670 doi: 10.17265/2161-6256/2015.08.004.

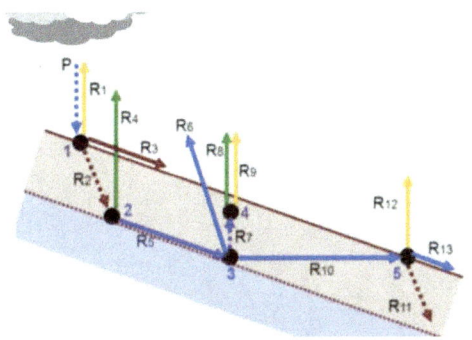

Fig. Thirteen secret water routes

It introduces a novel concept aimed at optimizing rainwater management, enhancing livelihoods, and bolstering climate resilience. Therefore, it is crucial to implement this approach at the grassroots level to demonstrate how resource accumulation can address water scarcity while concurrently mitigating climate challenges. The author trusts that upon digesting this book, you will find yourself in agreement. It is truly imperative to pinpoint the authentic formula for progress, especially in developing nations.

www.ingramcontent.com/pod-product-compliance
Lightning Source LLC
Chambersburg PA
CBHW070156230526
45471CB00002B/682